YOUR KNOWLEDGE HAS VALUE

- We will publish your bachelor's and
  master's thesis, essays and papers

- Your own eBook and book -
  sold worldwide in all relevant shops

- Earn money with each sale

Upload your text at www.GRIN.com
and publish for free

Bibliographic information published by the German National Library:

The German National Library lists this publication in the National Bibliography; detailed bibliographic data are available on the Internet at http://dnb.dnb.de .

Imprint:

Copyright © 2018 GRIN Verlag
Print and binding: Books on Demand GmbH, Norderstedt Germany
ISBN: 9783668741430

This book at GRIN:

https://www.grin.com/document/431389

William Fidler

Dark Matter. New Structures, Old Particles

GRIN Verlag

Dark Matter: New Structures, Old Particles.

W M Fidler

<u>Abstract</u>

It is shown that there may exist at least two different forms of dark matter fashioned from existing baryonic matter, specifically, neutrons. The structures developed are considerably more massive than the number of particles from which they are formed and are argued to be candidates for dark matter since they have no electrical charge, no electrons, hence no chemistry and no photonic emission. They can only interact with each other and other baryons through the agency of gravity. Further, one of the forms is much more likely to clump than the other and raises the possibility that there are large-scale formations of different types of the substance, one of which may more easily be disrupted than the other.

This may present further challenges for astronomical experimenters in the task of observing and identifying dark matter, for, on a large scale the distribution of dark matter within, for example, a galaxy may be unique to that constellation. In addition, detection of these structures on a small scale, by the usual method of collision with other matter, may be rendered difficult, if not impossible, for the structures can only be accelerated to high speeds by interacting with intense gravitational fields.

List of contents

Introduction

The concept of dark matter has its origin in the study by Zwicky in 1933 of the Coma cluster. He determined that the outlying galaxies in the cluster were moving substantially faster than the calculated velocities based upon the visible mass in the cluster and proposed that the cluster must be composed of over 90% of matter which could not be seen. He christened this matter, dark matter. His calculations were disparaged at the time, but modern estimates agree, fairly closely with this value.

Since the discovery of these anomalous velocities, observations, made by different methods, have established unequivocally that, throughout the universe dark matter is ubiquitous. A fairly comprehensive survey of the evidence for the existence of dark matter may be found in [1].

Over the years many attempts have been made to explain the nature of dark matter by physicists, principally by the invention of new particles, none of which has, to date been detected, despite heroic efforts so to do. A specific example of a new particle is the axion, favoured by Wilczek [2] and which had been devised for a problem not connected with dark matter. The non-appearance of these new particles has led to the suggestion that dark matter is not a new particle, but may be some exotic combination of ordinary matter. Some proposals for these combinations are reviewed briefly in the article by Hossenfelder & Lubick, [3].

It is in this spirit that we present in the following work, two combinations of the same particle of ordinary baryonic matter, namely, the neutron. It is shown that both structures have the property of substantially greater mass than the constituent particles in addition to all of the attributes of dark matter, viz; no electrical charge, no electrons and hence no chemistry and no photonic emission. The model of the vacuum used here is that explained at length in [4].

<u>**Frequency intervals**</u>

The model of the vacuum developed in [4] exists in two modes, one of which, the sterile mode is considered to be overwhelmingly dominant, in the temporal sense, in the evolution of the universe. The other mode, called the free vacuum mode exists only fleetingly but is of prime importance, for it is posited to be associated with particle formation and the various heatings of baryonic matter described in [4].

We proceed to examine the free vacuon mode to determine the frequency intervals over which the energy equivalents of the dark matter and baryonic matter are released into space.

These frequency intervals may viewed as proxies for time intervals, and, it is shown that, in the positioning of the frequency intervals close to the inception frequency of the universe it may be inferred that the time intervals for the formation of dark and baryonic matter from the energy released into space, were extremely small.

The energy, E_b associated with the total amount of baryonic matter in the universe is given by:

$$E_b = N_b m_b c^2 \text{ ----------------------- (1).}$$

In this formula N_b is the number of baryons in the universe, considered to be constant outwith the inception phase; the mass, m_b of a baryon is taken to be that of a proton, viz, $1.67252 * 10^{-27} \; kg/m^3$, although in view of the particle of which the structures are composed, then, pedantically, the mass of the neutron would be more appropriate.

Now, the model predicates that the above energy is directly associated with the liberation of free vacuons into space.

Thus, $E_b = N_{enc} \; 3h/2 \, (v_i - v_f)$ -------------------------------- (2).

The subscripts on v indicate the beginning 'i' of the release of free vacuons, whilst 'f' denotes the end of the phase.

It is shown in [4] that if we adopt Eddington's number of 10^{80} for the number of baryons in the universe, then the number of enclosures, N_{enc}, is $2.628 * 10^{92}$. From (1) and (2) we get:

$$(v_i - v_f)_b = 2N_b m_b c^2 / 3hN_{enc} \text{ -------------------- (3).}$$

Hence, the baryonic matter frequency interval, $(v_i - v_f)_b = 0.0576 * 10^{12} \; Hz.$

If we denote the total masses of dark and baryonic matter in the universe as M_d and M_b, respectively, then it follows from equation (2) that we may write:

$$M_d/M_b = \frac{(v_i - v_f)_d}{(v_i - v_f)_b} \quad\text{---------------------- (4).}$$

Outwith the dark and baryonic matter formation phases, which will be shown to occupy very short time intervals in the evolution of the universe, no more of these matters are made beyond the inception phase and hence the masses of dark and baryonic matter are thereafter constant in time. It then follows that the LHS of equation (4) is, throughout virtually the whole life of the universe, constant. Further, if we divide above and below by the volume, V_u, of the universe at any time the magnitude of the ratio is unchanged, but the LHS now becomes the ratio of the mass densities of dark, to baryonic, matter.

We may determine the magnitude of the density ratio from the results of the WMAP probe [5]. It was found that the current baryonic matter and dark matter mass densities, as proportions of the critical mass density were **4.6%** and **24%**, respectively.

There is no requirement to determine the actual densities in the modified version of equation (4), for it is plain that the ratio has magnitude, $24/4.6 = 5.217$. Hence, from equation (4), given that the baryonic frequency has already been established, the dark matter frequency interval is calculated to be:

$$(v_i - v_f)_d = 5.217 * 0.0576 * 10^{12} = 0.3 * 10^{12} \; Hz.$$

The shape of nucleons and the quark cylinder.

The Standard Model of particle physics regards the nucleons to be of spherical shape. In the case, specifically of the proton and neutron, it is more correct to say that the quarks, of which these particles are composed are constrained within a spherical envelope.

In keeping with the assertions in [4] we posit that during the initial stage of the dark matter frequency interval, which we will show should be positioned at the inception of the universe, baryonic matter, wholly in the form of atomic hydrogen was produced. It is in the extreme compression of this matter that it is converted completely into neutrons which are subsequently deformed into the elements of the structures proposed for dark matter.

Montgomery and Jeffrey [6], propose that the nucleons are not spherical but exist in the form of triangular ovoids with a quark at each vertex. Whilst their work is principally concerned with the structure of the atomic nucleus and attributes structure to the arrangements of the electrical charges of the quarks, it is extended here to describe possible candidates for dark matter.

Given that the initial shape of the neutron is that of a triangular ovoid, then, under the extreme conditions at the inception of the universe we posit that, in the limit, the triangular ovoid may be deformed into a plane equilateral triangle, as shown in Fig1.

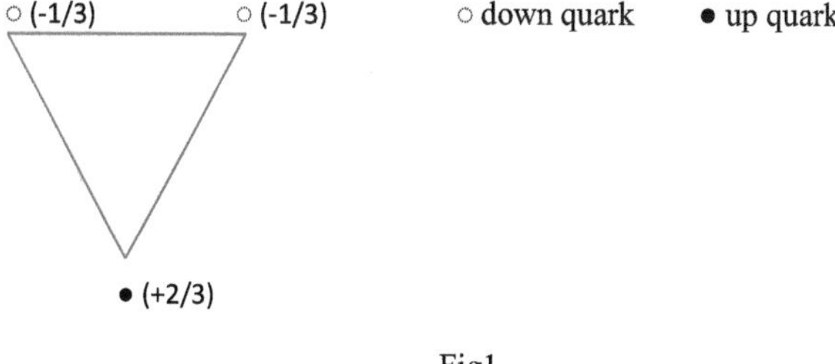

Fig1

It may be noted that the quarks are located inside the triangle and at the vertices, but are as depicted for the purpose of clarity. If electrical forces are the principal agents of action between neutrons, then the above configuration may be used to show why neutron-neutron pairing does not occur, despite the nucleon having no net charge. The reader is invited to superimpose a copy of the above on the figure. It will be found that there is no orientation where the charges are all attractive. This argument is used by Montgomery and Jeffrey to show that all pairings of neutrons in this superimposed configuration are unstable.

Similar arguments may be used to show that proton-proton pairings are also unstable.

If, however we combine a proton and a neutron it is seen that there is one configuration where the charges are all attractive. Of course, such a combination is the deuteron, which is known to be stable.

As justification for the deformation of the ovoid into a triangle we refer to the work of Llanes-Estrada and Navarro, [7]; there it is shown that under the extreme pressures characteristic of those which may be found within neutron stars, the shape of neutrons can change progressively from spheres into cubes. However, it was argued that the pressures required were considered too large for the neutron stars then observed, for the masses of such stars were in the region of 1.4 solar masses and the pressures generated within were too small to produce cubic neutrons. Since stars of mass larger than approximately two solar masses collapse to form black holes, it was thought that there were no neutron stars in which cubic neutrons could form. However, in 2010 a neutron star, labelled PSR J1614-2230, in the constellation of Scorpius, was discovered with a mass, measured to be 1.97 solar masses. This is close to the boundary between neutron stars and black holes and calculations showed that it was massive enough to generate pressures which would permit the formation of cubic neutrons.

We now form a combination of neutrons which coincide but lie in the same plane, as shown in Fig2.

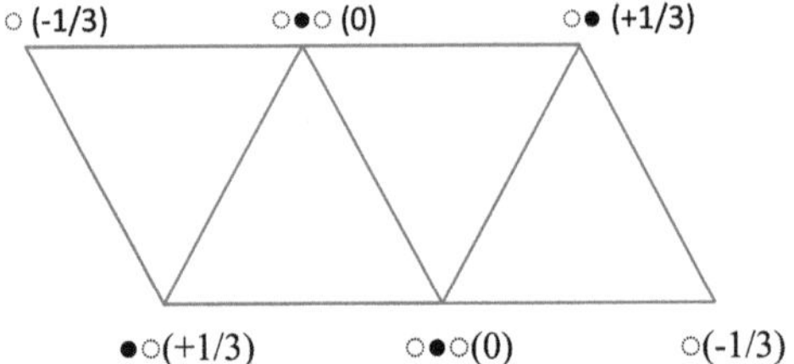

Fig2

The above configuration may be rolled up into a closed band and then, at the confluence of all of the vertices we have the quarks combination ○●○. This taken to be a combination of maximum attraction, whilst at the same time has zero charge. Hence, we have a structure which has no electrical charge, no electrons, hence no chemistry and cannot emit photons. The structure can only interact with other structures of the same or different types, through the agency of gravity. Further, we posit that the dark matter remains in a cold state for it is present during the baryonic matter phase, in which we show that the energy liberated during this phase is only sufficient to form baryons; the dark matter does not interact with the energy liberated into space during this phase, and is then considered to be thermally inert.

It is of interest to note that a similar structure cannot be formed with an odd number of neutrons unless it is formed into a Mobius band. Whilst such is entirely possible mathematically, it is considered to be only a remote possibility and will not be discussed further.

The combination shown in Fig2 may be extended indefinitely, but, as the extremities become further away from each other, the possibility of the rolling up into a band becomes less likely. It is obvious that the energy configuration of the bands is directly related to the number of neutrons in the structure. It then follows that the combination having the lowest energy contains only two neutrons; the rolled up version, which has the shape of a cylinder, is proposed as a candidate for dark matter; such an arrangement is shown in plan view in Fig3.

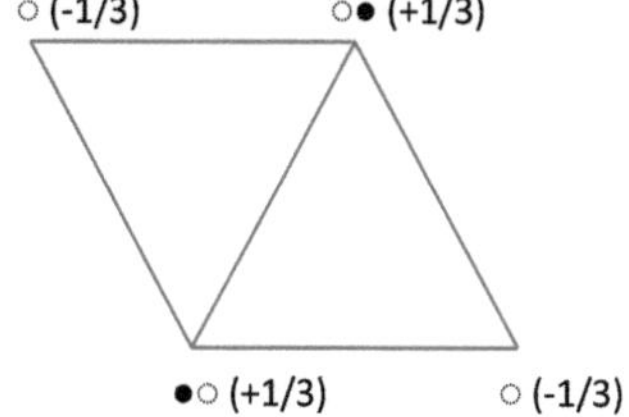

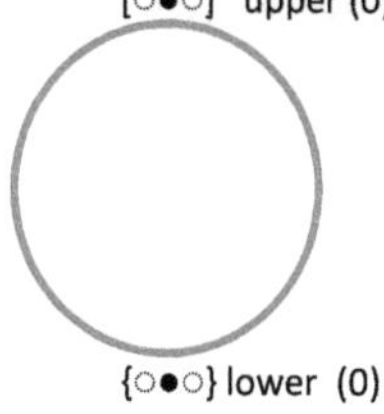

Fig3

The quark crystal

We name the following structure the quark crystal for it has a passing resemblance to the structure of a conventional crystal. Consider the configuration of neutrons and the corresponding quarks arrangements shown in Figs4&5, below.

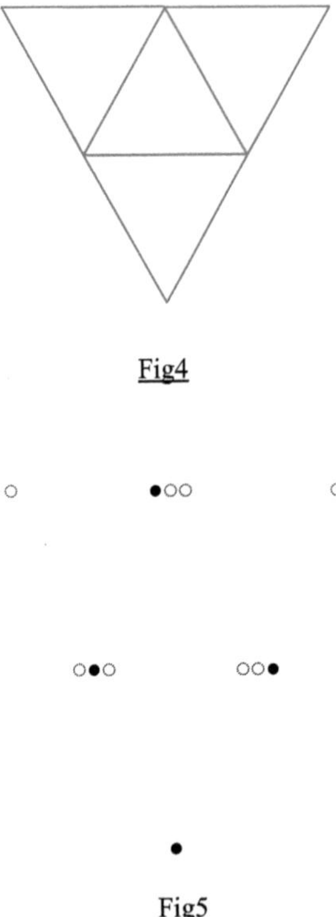

Fig4

Fig5

If the 'exterior' triangles of Fig4 are rotated about the sides of the 'inner' triangle in a direction up out of the plane of the paper until the outer vertices meet, then the structure thus formed has the shape of a triangular pyramid, as shown in Fig6.

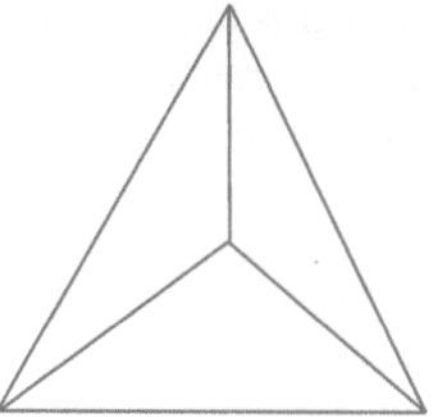

Fig6

If we examine this structure it may be seen that there is no net charge at the confluence of the vertices; as may be inferred, the structure possesses the same other properties as the quark cylinder, and as noted previously, resembles a crystal, hence the name chosen for the structure.

For convenience we now assume that the side length of the neutron which is deformed into a triangle is the same size as that of the diameter of a spherical neutron, viz: approximately 10^{-15} m. We may now determine the volume of the quark cylinder and the quark crystal.

It is easy to show then that the volume of the quark cylinder, v_r and the quark crystal, v_c are given by the formulae: $v_r = \frac{\sqrt{3} * 10^{-45}}{8\pi}$, and $v_c = \frac{10^{-45}}{6}$, respectively.

The frequency of the vacuum at the inception of the universe may be calculated if we consider all of space to be filled with either one or other of the shapes above.

Taking the Eddington number as N_b and the number of deformed neutrons in a structure as n_s, then, in general, $\frac{N_b * v_i}{n_s} = N_{enc} * l^3$, where l is the side of an enclosure. Now,

$$vl = c$$

Hence we may write: $v_i = \sqrt[3]{\frac{n_s * N_{enc}}{N_b} \frac{c^3}{v_i}}$ ---------------------- (5).

For the quark cylinder case ($n_s = 2$), $v = 1.28 * 10^{28}$ Hz.

For the quark crystal case ($n_s = 4$), $v = 1.202 * 10^{28}$ Hz.

Since we have proposed that the dark matter is formed from baryonic matter, then it follows that baryonic matter was formed first at the inception of the universe; dark matter was formed from this matter, and thereafter, the baryonic matter that may be observed currently in the universe was formed.

It is then plain, given the magnitudes of the frequency intervals calculated previously, that in comparison with either of the frequencies of the vacuum calculated above, the frequency drops associated with the free vacuon mode (which is assumed to extend over the sum of the two frequency intervals) are indicative that both dark and baryonic matter formation occurred in very short intervals of time.

It has been posited in [4] that matter is formed from the intense curvature of space locally causing closure of space and trapping energy of magnitude, $E = mc^2$ within. Here, **m** is the mass of the particle. On this basis, for example, the energy trapped within the particle called a proton is $1.505*10^{-10}$ **J** .

Now, it is to be expected that space will be curved intensively if the rate at which energy is being liberated into space vastly exceeds the rate at which space is being created.

We now proceed to give proof of the statement made in [4] that such a ratio is given by the ground state energy density.

In the model developed in [4], the minimum energy that may be released into space is the vacuon, of magnitude, $= \frac{3h\Delta}{2}$.

It then follows that, in finite difference form, the rate at which energy is being deposited in space is given by $\frac{\delta E}{\delta t} = \dot{E} = \frac{3h\Delta}{2\delta t}$ ------------------------ (6)

Now the volume of an enclosure, $V = l^3$, hence $\delta V = (l + \delta l)^3 - l^3$. If δl is very small then we may write, $\delta V = l^3\left(1 + 3\frac{\delta l}{l}\right) - l^3 = l^3\,\frac{3\delta l}{l}$.

The ground state energy density is given by $\rho_G = \frac{h(v - \Delta)}{2l^3}$.

Hence, $\rho_G \delta V = \frac{3h(v - \Delta)}{2} * \frac{\delta l}{l}$.

Now, both of the products, vl and $(v - \Delta)(l + \delta l)$ are equal to the speed of light, **c**. From the second of these we may write: $vl.\left(1 + \frac{\delta l}{l}\right).\left(1 - \frac{\delta}{v}\right) = c.$

It may then be shown that $\frac{\delta l}{l} = \frac{\Delta}{(v - \Delta)}$.

Hence, the product, $\rho_G \delta V = {}^{3h(v - \Delta)}/_2 * {}^{\delta l}/_l = {}^{3h\Delta}/_2$.

We may write that ${}^{\delta V}/_{\delta t} = \dot{V} = {}^{3h\Delta}/_{2\rho_g}$ ------------------------ (7).

Dividing (6) by (7) produces the result claimed, viz: ${}^{\dot{E}}/_{\dot{V}} = \rho_G$ ------------- (8).

It is of note that the expression shown above for the ground state energy density contains the factor l^3 in the denominator and not, $(l + \delta l)^3$. This is justified because, in the model used here the oscillator must first fall from the first state energy level to the ground state, at which the expansion of space has not taken place due to the inhibiting effect of the vacuon liberated. As explained in [4] the expansion of space occurs after the oscillator falls into the ground state.

<u>**The equivalent masses of the cylinder and crystal structures.**</u>

Since the elements of which the structures are formed have been deformed from their 'normal' shape, then the effective mass of either structure must take account of the energy stored within the deformed elements.

If it is assumed that the stored energy in either structure is some proportion, **y** of the of the energy equivalent of the baryonic matter, then, in general, we may write:

$$M_d c^2 = {}^{N_b}\!/_{n_s}\, y.\, m_b\, c^2 + N_b m_b\, c^2$$

This may be written: $M_d = {}^{M_b}\!/_{n_s}\, y + M_b$, from which we obtain:

$$^{y}\!/_{n_s} + 1 = {}^{\rho_d}\!/_{\rho_b} = 5.217. \text{----------------- (9).}$$

For the cylindrical quark structure, $= 8.434$; for the crystal quark structure, $y = 16.868$.

Hence, in multiples of the mass of a baryon we find that the effective mass of the cylinder is **10.434,** whilst that for the crystal is **20.868**. These structures are formed from two and four neutrons, respectively.

Discussion

We have shown that there are at least two structures considerably more massive than their baryon constituents (specifically, neutrons). Using the magnitudes calculated for the frequencies of the oscillators at the times at which the total number of these structures were formed, and the frequency intervals previously determined, we may state with a degree of confidence that these structures were formed at the inception of the universe, and further, using either of the frequencies calculated previously, it may be inferred that matter formation, attributed to the closing of space about energy will occur at the time at which the ground state energy density of the oscillators is a maximum; a simple calculation will show that at the inception of the universe, where the frequency of the oscillators is of the order of $10^{28}\ Hz$, the ground state energy density is of the order of $10^{49}\ {}^{J}\!/\!_{m^3}$. It should be noted that this

may be interpreted as the rate at which energy is entering space relative to the rate at which space is being created.

We have not explored the possibility that both types of structure are produced simultaneously, but given the closeness of the frequencies calculated, it is distinctly probable.

This introduces an interesting situation in that, due to their shape, clumping of the cylinders may not be particularly widespread within a body of such objects, whilst even a cursory inspection of the shape of the crystal shows that clumping is very likely. Hence, it may be inferred that at least two different large-scale structures of dark matter are extant. This will only compound further the difficulty of detection.

W M Fidler, June 2018.

References

[1] The Whole Shebang.

Timothy Ferris, 1997.

Weidenfeld & Nicholson

The Orion Publishing Group.

[2] The Lightness of Being.

Frank Wilczek, 2008

Penguin Group.

[3] Strangely Familiar

S Hossenfelder & N Lubick

New Scientist, 22 August 2015, pp28-31.

[4] Cosmological Heresies, a new model of the evolution of the universe.

W M Fidler,

ISBN 9783668652224

GRIN Verlag.

[5] WMAP-Content of the universe.

http://map.gsvc.nasa.gov/universe/uni_matter.html

[6] Lattice/modified cluster model of the atomic nucleus.

J Montgomery & R Jeffrey, 2008

Unclear2nuclear.

[7] Neutrons become cubes inside neutron stars.

F Llanes-Estrada & G M Moreno

arxiv.org/abs/1108.1859.

MIT Technology Review, Aug 2011.